The First Atom & Center of the Universe

Ecophysics Practitioner
Mohammad Abu Shahadat

The First Atom & Center of the Universe

by

Mohammad Abu Shahadat

Contents

Serial no	Tittle	Page
1.	About the Author	4
2.	Author Speech for the Reader	5
3.	First Atom of the Universe	6
3.1.	Center Point Theory	8
3.2.	Calculation of Radius and Area of the Universe	13
4.	Location of Angels, Aliens and Animals	23
4.1.	Angels	23
4.2.	Aliens	24
4.3.	Animals	29
5.	Theory of Travelling Time in the Universe	30
6.	Theory of the Global Warming	34
6.1.	Fundamental Concept of Global Warming Model	35
6.2.	Global Warming Micro Model	36
6.3.	Global Warming Macro Model	44
6.4.	Some Function for Global Warming	47
6.5.	Who are responsible for Global Warming	51
6.6.	How to Reduce the Global Warming	52
6.7.	Turning the Present World into the Green World	56

About The Author

Mohammad Abu Shahadat

Founder & Headteacher, Notre Damian School
Student, Welfare Economic System since 2013
Practitioner, Ecophysics
BSS (Hon's), MSS in (Economics), Jagannath University
Studied from 2002 to 2004 at Notre Dame College,
from 1997 to 2002 at Delpara High School and
from 1992 to 1996 at MDC Model Institute
Born on 14 January 1987, Dhaka, Bangladesh

https://www.amazon.com/Mohammad-Abu-Shahadat/e/B07WC912RD

Contact: www.facebook.com/MAbuShahadat
Email: mabu.shahadat@gmail.com
Cell Phone: +8801682744844

The Author's Speech for The Readers

I founded Welfare Economic System and Development in 2013. Since then to develop the welfare economic system I individually have been working and researching. Without the mystery of the universe and global warming solution I cannot complete the Welfarism.

Dear readers and followers, I am none without you. I always try to write different topics. I don't disappoint you to get new concepts and thoughts. As no scientist or researcher has not still discovered the location of alien, the mystery of graviton particles, center of the universe and first atom. So I have chosen these topics to write my thoughts with my little knowledge. So if you find any mistake please inform me by email. First Atom & Center of the Universe is my 4th short book.

My wife and I are trying to educate our students with qualified teachers. Students of our school come from various economic groups but they get equal opportunities to learn. We hope, we will create future Newton, Einstein, Socrates, Pythagoras and so on. Thanks for reading this section.

Mohammad Abu Shahadat
21 September 2019
Dhaka, Bangladesh

The First Atom of the Universe

We know that according to the Big Bang concept, there was nothing except an super atom in the universe. Then anyhow the super atom was blasted and since then the observable universe is expanding. The observable universe is 13.5 billion years old and the earth is 4.5 billion years old. So the universe is three times older than the earth. But recently NASA has found a star which is about 16 billion years old. By this star's age it is proved that the creation time of visible universe is not true according to the Big Bang Concept. Suppose, the universe is a father and the recent finding star, Methusela or HD 140283, is a son. As a son can never be born before a father and everything has a center, I have firmly said that the Big Bang is a concept not a theory. I think, the universe is a big circle like our eye ball. As every circle has a center. So I firmly believe the Universe has a center. The universe is moving like a fan. All of the things has the definite uniform circular motion but different speed. Some think the sun is the center of our solar system. Some think the black hole of our milky way is the center. Some think that there is no center of the universe. There are more than one black holes in the universe and they are stars without shining power but super gravity. I think any star cannot be center of the universe. When I am trying to develop the Global Warming Macro Model a question came in my mind "Which is the center of the Universe?" I thought several times and watched nearly all videos about the universe on Youtube Channel and read various articles but could not find any solution. Scientists, researchers, students of Astronomy believe that there is no center of the universe.

However, I have to find out where the first atom is. I think this is the point from where all the things of the universe get power to move their way like our navel point. Everything of the universe is dynamic. Some move round black holes. As black holes are a result of dying stars, black hole is not center. All things including black holes get circular power from the central point.

Now the question comes , " **How did the visible universe create ?"**

The Creator of the Universe knows well how He creates the universe. From the ancient time men have given many concepts about the universe. Of them the big bang is the most popular concept.

I think, once there was nothing except a super atom in the universe. This particle is very hot red colored. It moved very fast like a top. By spinning the atom created smoke and for spinning it was displacing slowly and created its own uniform circular axis way. This circular axis way is the center way of the universe. The smoke was gradually expanding and scattering various places in the universe. By gravity smoke of various places created planets, sub-planets, stars, comets, galaxies, local group galaxies etc. Everything has own uniform circular axis way for gravity. Dark matter creates gravity. Graviton particles are invisible. Various elements were concentrating the super atom slowly. As a result the atom was losing its flaming power gradually and becoming cool. These elements including the super atom created a planet. The interior part of this planet is very hot and the outer part has water, soil surrounding the super atom. The ratio of water and soil is about 2.45 : 1 . Of the whole water the purest drinkable water has near the super atom. This planet has CO2O2 cycle, a sub planet, animals, plants and aliens (visible and invisible). This is the only one planet whose has animals and aliens.

Center point theory

If the power of center point is zero, every thing of the universe will lose their circular and dynamic power and then universe is static that means time is zero. If it lasts few minutes all things lose their gravity, they will collide one another. The moon will lose its shining. As a result all the things of the area of the universe will destroy.

The Universe moves around the Holy First Particle of the Universe. Some think there is no center point in the universe. But think, without a central point how a circle is made. So without a center we cannot think about the universe. Stars, Solar system, star clusters, Galaxies, Galaxies clusters all have uniform circular motion. They move around the Holy First Atom. The universe has a center axis way. By this axis way a planet is moving. The planet has CO2O2 cycle and living beings. The Holy First Atom lies in the middle of the planet.

"The Universe is expanding "is a wrong idea. The things of the universe have a definite shape and a definite circular way and the dynamic power that comes from the central point of the Universe. To maintain balance they are dynamic and run their definite uniform circular axis. For this we see some planets or comets after every a long time. For example, Halley'scomet appears in every 75 years. As everything of the universe is moving it seems to expand the universe. But really the universe is no expanding rather the number of the things is unchangeable.

The things can come to close one another according to their definite axis. They can not come out from their definite axis. If all things meet one another they will be a straight line. And the central point of the universe is in a middle position. Then we get the diameter of the universe. The diameter of the Universe is 254r billion light years. Here r is equal to 46.5 billion light years approximately. The angle of the straight line is 180°. But they do not collide one another because of wide space among them.

All the things of the universe move around the central point according to their own uniform circular motions. Central point is in the middle of a planet. This planet has CO_2O_2 cycle, animals and aliens. If the charging power of the central point is zero level, the universe is constant. As a result, there will be no motion in the universe. Surprisingly the central point gets charging power by human beings and invisible aliens.

Let's find out the first atom of the Universe. As every word of the Holy Quran is true, so we have to depend on the holy Quran. According to the Holy Quran the Universe consists of seven havens (skies) and made it in 6 cosmic days and also made the earth 2 cosmic days.

6 cosmic days / 2 cosmic days = 3

The verse number of Sura Al- Isra(The night journey) is 111

There are 3 pillars in the Kaba Sharif. The Kaba Sharif is the holiest place in the visible universe. The 3 pillars are situated just like 111. The 1st verse and 44th verse of the Sura Al- Isra are the symbol of the measurement, size ,position and center of the universe I think. Moreover, the 12th verse of this Sura is the symbol of the calculation of time, day, month and year.

1 cosmic day = 50000 years of the earth [According to the 4th verse of 70 Sura Mayariz]

86400 seconds of cosmic

= 50000 X 365 X 86400 seconds of the earth

1 cosmic second

= 50000 X 365 seconds of the earth

= 18250000 seconds of the earth

1 cosmic second = 211.227 days of the earth

1 cosmic second = 5470054250000 km

[since, speed of light = 299729 km / second]

1 cosmic second = $5.47005425 \times 10^{12}$ km

We know that ,

1 second of the earth = 299729 km

$$= 2.99729 \times 10^5 \text{ km}$$

The difference between the cosmic second and earth second = 5470054250000 km - 299729 km

$$= 5470053950271 \text{ km}$$

There are two special speeds in the universe. They are the speed of light and the speed of angel.

The speed of light = the speed of alien (Jinn)

= 299729 km per second

The speed of angel = 800 light year per second [Approx.]

The HOLY POINT is in the middle point of the Kaba Sharif which is at Mecca in Saudia Arabia. In the Kaba Sharif we see that there are three pillars. The Middle pillar is the center of the universe. The first atom of the universe is in this pillar. The big bang's point which is blasted for making the universe is there. Other two pillars has equidistant from the middle point. Man and Jinn move seven times around the Kaba Sharif. That means the universe is charged by human beings and jinn according to the Creator directions. As a result, all the things get charged to move their definite axis. If there is no man to move around the Kaba Sharif then the Universe will lose its circular motion power and the Universe will destroy. The day no man finds out to move the Kaba Sharif, that day the whole universe will be destroyed.

According to the Holy Quran, the beginning point (Uc) of the Universe is the middle pillar of the Kaba Sharif and the last point (U_L) of the Universe is Sidratul Muntaha. Kaba sharif has three pillars (1 - 1 - 1) . The verse number of Sura Al-Isra (The Night Journey) is 111.The Muslim have to move Seven times around the Kaba Sharif. As the Universe has seven heavens, the Muslim have to move seven times around the Kaba Sharif. I think, by moving 7 times around The Kaba Sharif human being charges the central point of the universe. This power goes all the things of the universe to get uniform circle motion. Rainbow has 7 colours and we know that white is made by 7 colours. The muslim when move around the Kaba Sharif they wear white coloured clothes. I think seven heavens, seven colours of rainbow and white colour of the Hazis' clothes have a connection.

1 and 44 verse of Sura Al-Isra (The Night Journey) of the Holy Quran are the source of my thought.

1 . " Glory to Him who journeyed his servant by night, from the Sacred Mosque to the Farthest Mosque, whose precincts We have blessed, in order to show him of Our wonders. He is the Listener, the Beholder."

Here, The Sacred Mosque means the Kaba Sharif. As the journey started from the Kaba Sharif, the center of the universe is the Kaba Sharif.

44. " Praising Him are the seven havens, and the earth and everyone . There is not a thing that does not glorify Him with praise, but you do not understand their praises. He is indeed Forbearing and Forgiving "

"The seven havens and the earth " is clue to measure the radius of the Universe. The center point is the middle pillar of Kaba Sharif. We see that the people who come to Makka in Soudi Arabia to say prayer all are moved seven times around the Kaba Sharif.

We cannot see the Creator but we believe and feel His existence everywhere. The Creator is One. He runs the whole universe.

Scientists cannot see the graviton particle but they believe that graviton particles have in the universe.

Therefore, I think the universe is not expanding , it is a circle and it has a central point and central axis. All the things of the Universe are moving around the middle pillar of the Kaba Sharif according to the uniform circular motion.

Calculation of the Radius and Area of the Universe

What we see with open eyes and with telescope is in the first sky (haven). They are a very little part of the radius of the first heaven. Stars, Solar system, Nebulae, star clusters, Galaxies, Galaxies clusters , black holes etc. are in the first circle . Consecutively diameter of one circle is the radius of other circle. That is, If the radius of the first circle is r , the diameter of the first circle (First haven) will be $2r$. The diameter of the first circle is the radius of the second circle (the second). Continuously the radius of the seventh haven circle is the diameter of the sixth haven.

The radius of the Universe = 127r

$$r(1+2+4+8+16+32+64) = 127r$$

U_C ————————————————————————— U_L

U_C is the central point and U_L is the last point of the universe

r = about 46.3 billion light years
The sum of seven circle's radius
= $(2^0 + 2^1 + 2^2 + 2^3 + 2^4 + 2^5 + 2^6)\, r$ billion light years
= $127\, r$ billion light years

The circumference of the Universe
= $254\, \pi\, r$ billion light years

The area of the Universe
= $\pi\, (127\, r)^2$ square billion light years
= $16129 \pi r^2$ square billion light years
= $16129 \times 3.1416 \times (46.5)^2$ square billion light years
= $16129 \times 3.1416 \times 2162.25$ square billion light years [approx.]
= 109563080.8734 square billion light years [approx.]

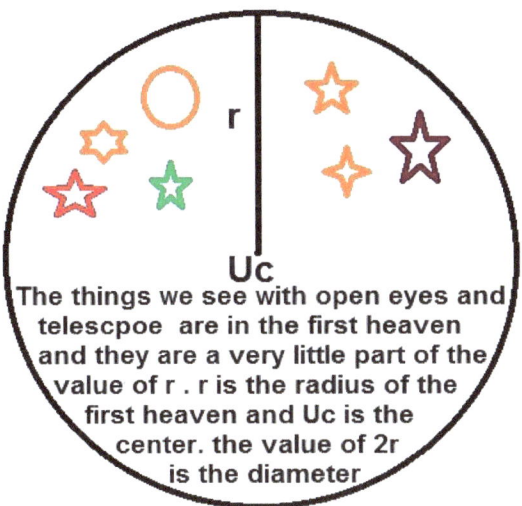

Figure: The First Heaven

r is the radius of the first heaven
Here, r = 46.5 billion light years [approx.]
Π = 3.1416

The diameter of the first heaven
= 2r
= (2 X 46.5) billion light years [approx.]
= 93 billion light years [approx.]

The circumference of the First Heaven
= 2 Π r
= 2 X 3.1416 X 46.5 billion light years [approx.]
= 292. 1688 billion light years [approx.]

The area of the first Heaven
= πr^2
= 3.1416 X (46.5)² square billion light years
= 3.1416 X 2162.25 square billion light years
= 6792.9246 square billion light years

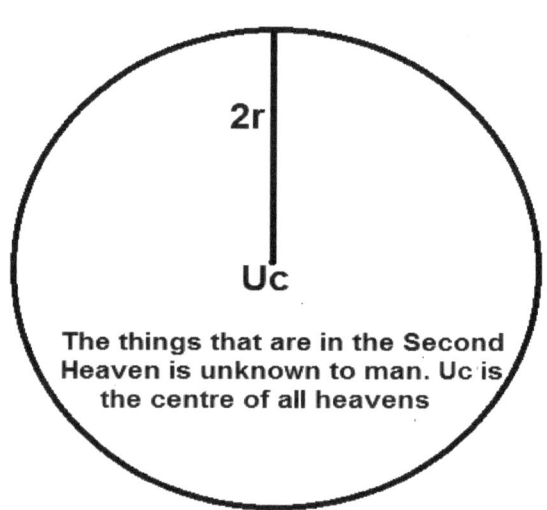

Figure: The Second Heaven

The circumference of the Second Heaven
= 4 π r billion light years
= 4 X 3.1416 x 46.5 billion light years
= 584.3376 billion light years
The area of the Second Heaven
= $\pi (2r)^2$ square billion light years
= $4\pi r^2$ square billion light years

= 4 X 3.1416 X 2162.25 square billion light years
= 27171.6984 square billion light years

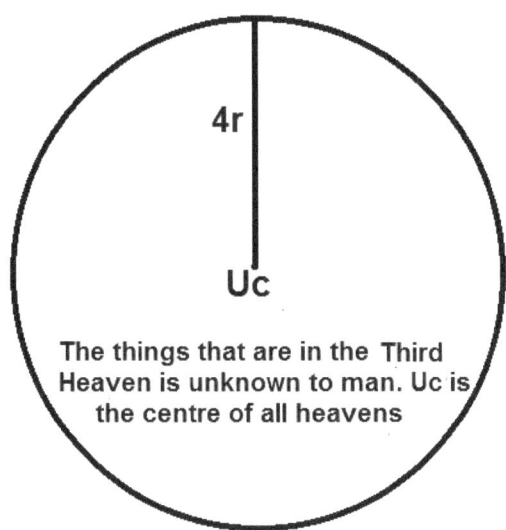

Figure: The Third Heaven

The radius of the third Heaven

= 4r = 4 X 46.5 billion light years

= 186 billion light years

The circumference of the Third Heaven
= 8 π r billion light years
= 8 X 3.1416 X 46.5 billion light years
= 1168.6752 billion light years
The area of the third Heaven
= $\pi (4r)^2$ square billion light years
= $16\pi r^2$ square billion light years
= 16 X 3.1416 X 2162.25 square billion light years

= 108686.7936 square billion light years

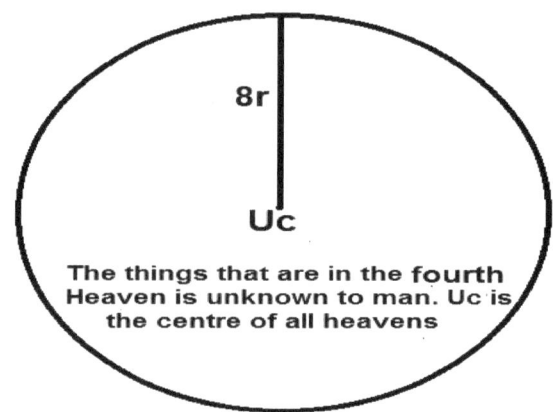

Figure: The Fourth Heaven

The radius of the fourth heaven

= 8r = 8 X 46.5 billion light years

= 372 billion light years

The circumference of the fourth Heaven
= 16 π r billion light years
= 16 X 3.1416 X 46.5 billion light years
= 2337.3504 billion light years

The area of the fourth Heaven
= $\pi (8r)^2$ square billion light years
= $64\pi r^2$ square billion light years
= 64 X 3.1416 X 2162.25 square billion light years
= 434747.1744 square billion light years

Figure: The 5th Heaven

The radius of the 5th heaven
= 16r
= 16 X 46.5 billion light years
= 744 billion light years

The circumference of the 5th Heaven
= 32 ∏ r billion light years
= 32 X 3.1416 X 46.5 billion light years
= 4674.7008 billion light years

The area of the 5th Heaven
= ∏ (16r)² square billion light years
= 256∏r² square billion light years
= 256 X 3.1416 X 2162.25 square billion light years
= 1738988.6976 square billion light years

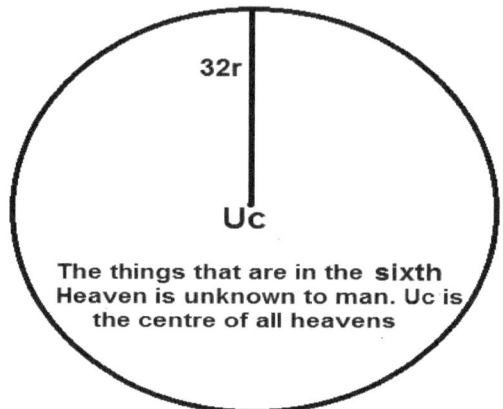

Figure: The Sixth Heaven

32r is the radius of the sixth heaven

The circumference of the sixth Heaven
= 64 π r billion light years
= 64 X 3.1416 X 46.5 billion light years
= 9349.4016 billion light years

The area of the sixth Heaven
= π $(32r)^2$ square billion light years
= 1024πr^2 square billion light years
= 1024 X 3.1416 X 2162.25 square billion light years
= 6955954.7904 square billion light years

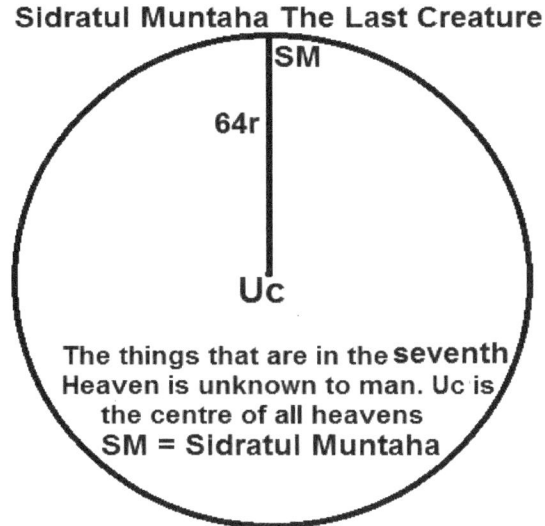

Figure: **The Seventh Heaven**

The radius of the seventh heaven
= 64r
= 64 X 46.5 billion light years
= 2976 billion light years

The circumference of the seventh Heaven
= 128 ∏ r billion light years
= 128 X 3.1416 X 46.5 billion light years
= 18698.8032 billion light years

The area of the seventh Heaven
= ∏ (64r)² square billion light years
= 4096∏r² square billion light years
= 4096 X 3.1416 X 2162.25 square billion light years
= 27823819.1616 square billion light years

The sum of seven circle's radius
= $(2^0 + 2^1 + 2^2 + 2^3 + 2^4 + 2^5 + 2^6)$ r billion light years
= 127 r billion light years
= 127 X 46.5 billion light years [Approximately]
= 5905.5 billion light years [Approximately]
So the Radius of the Universe is almost 5905.5 billion light years [Approximately].

The area of the universe
= $\pi (127r)^2$ square billion light years
= $16129 \pi r^2$ square billion light years
= 16129 X 3.1416 X 2162.25 square billion light years
= 109563080.8734 square billion light years [Approx.]

Position of the Creator

= $(16129 \pi r^2 + X^2)$ square billion light years

= $(109563080.8734 + X^2)$ square billion light years [approx.]

All the things of the area of $16129 \pi r^2$ square billion light years (approximately) will be destroyed on the last day of the Universe by the One who occurred the big bang .

When the charging power of the central point is Zero. All the things in the area of the Universe will destroy. Only The Creator is and will be remain everywhere.

Location of Angels, Aliens and Animals

Nowadays the word Alien is the most talked of topic in the world. Children are very interested to know about alien instead of ghost or fairy. Space crafts are sent one planet to another planet for finding out Alien. Still now science has failed to discover alien. Some scientists believe that there is life beyond the earth. For this they send rockets to another planet from the earth. Still now they have failed to discover the location of Alien. The picture of alien we see is not true. It is an unreal image. I think aliens live with us since the first man. There are two speedy creatures in the universe. Their speeds are faster than man-made vehicle's speed. human being can not see them. These two creature of the universe are angels and aliens. Here I would like to discuss about angels, aliens and animals and theory of travelling time in the universe and the fastest speed particle in the universe which is more speed than the light speed.

Angels

The graviton particles of the universe are not seen but we feel that there is a dark energy between two planets. Same way angels are not seen but we feel their existence invisibly. The other name of graviton particles is angel's particle. Dark matter and dark energy related to angles. Angels live everywhere in the universe. They can change their physical form. They are visible but due to the fastest speed in the Universe we , human beings, can not see them. Angel's speed per second is the distance that from the earth to the Rigel [approximately]. The distance from the earth to the Rigel is 800 light years. So Angel speed is about 800 light years per second. Their actual size and form is so gigantic that we cannot imagine it. Angles have many wings which help them fly faster than most other speed medias. The honest and innocent great men can see them.

Luckily, in the seventh century the most honest and innocent great man of the earth travelled the Universe with angle's speed one night. Duration of his journey was only a small part of that night. It is possible because of the help of angel speed. From his journey we learn that the last creature of the universe is The Sidratul Muntaha. The Sidratul Muntaha is a kind of tree.

Aliens

This is a concept that alien may exist beyond the earth. But we don't know that aliens live with us from the first human being. The modern name of Jinn is Alien. Jinn have two wings and move one place to another at the light speed which is 299729 km per second , invisible and sometimes visible but holding various forms. Most of the jinn come to human beings and live together invisibly. They can change their visible shape like as a chameleon changes its colour. There are two kinds of alien in the universe. One type is related jinn .The worst jinn and his fellow eat animals' blood and things of dustbin. They like to change original shape into snakes. The worst jinn or alien whom we call the devil lives in the middle of the Bermuda Triangle. The place is in the western part of the

Atlantic Ocean. The top point of the Bermuda Triangle is Bermuda and other two points are Miami and Puerto Rico. From there the worst jinn connects its fellow jinns and the successors of the worst jinn. Most of the successors are involved in homosexual, sucking human blood, creating quarrelling among human beings. Bad jinns love to the individuals who practice gambling, suicide, homosexuality etc.

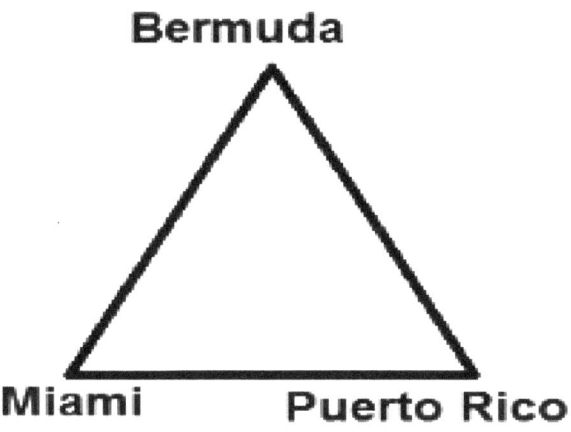

On the other hand, the best alien and his fellow live in the Jinn hill of Saudi Arabia and various places where people normally do not go or stay. Female jinn is called fairy. Jinn can travel any place with the speed of light. Sometimes we hear that someone of us has seen the Jinn and Fairy. Someone of us brings them up. They travel in the universe at night. From the earth they reach the moon within 1.28

second. Every night they travel one planet to other planet within a few moments. The honest jinn fellow love to eat sweets.

Another type is human-formed alien. Human–formed alien's fore parents were the worst Jinn and an Egyptian woman. Once human-formed aliens ruled Egypt. They were called Pharaoh. They are visible but the worst Jinn and his fellow jinn can not be seen.
Human-formed aliens are identified very easy. They love Pyramid because their predecessors were buried there and like to show sign of triangle various motions, one eye in a triangle. About 6000 years ago, in Egypt this type of aliens were born and gradually live with human beings. One night when the worst jinn, the Satan, travelled in the universe, he saw an unclean woman in Egypt. In guise of human being he came to the woman and stayed the whole night with her. After ten months she gave birth the first human-formed alien. When the son was adult, he would rape, lend gold with interest, do homosexuality, make people fool by telling lies and cheat for capturing his interest. Very often the worst jinn came to that man. Son of the worst jinn, the first human-formed alien, became the father of many human-formed aliens. He made friendship with the royal officials to meet the king of Egypt. Several times he lent gold to the royal officials to manage them in his need. Managing all of the royal officials once he killed the king and declared the king of Egypt . All royal officials obeyed to him because every royal officials borrowed a lot of gold to him. They carried out human-formed alien lest they should repay the loan

with high interest. Gradually the first human-formed successors spread their characters all over the world. At first they spread loan business. Gradually Banking system established.

I think , at present about 10 billion human-formed aliens live all over the world especially Israel, the USA, Australia, England, Germany, France, Austria, Japan, Chine, Mongolia, Brazil, India etc. For the conspiracy of human-formed alien the French revolution and Russian revolution occurred, British East India Company are forced to collect tax from the people of sub continental, the world war I and II happened, banking system spread and many political leaders of the world were killed. Nowadays most of the successors of the first human-formed alien's blood line live in the US. I respect John F Kennedy, the 35th President of the US, who wanted to free The US from human-formed aliens. But he was killed by the helper of aliens. Maybe the 41st and 43rd Presidents of the US are human-formed aliens. Because of their wrong decisions, people all over the world had been suffering many economical problems.

If the 43rd President, George W. Bush's DNA is tested and checked with the dead body of the Pharaoh which is in the Cairo Museum, it will be cleared whether Mr. Bush is alien or not. Only for establishing NEW WORLD ORDER the world trade center was damaged on 11 September 2001 through the conspiracy of alien Mr. Bush. Only for oil he attacked Iraq on the basis of false news. For Mr. Bush's guilty many innocent of Iraqi people were killed, raped and tortured. Only for Mr. Bush's wrong decision Euro crisis occurred and Spain, France, Greece faced economical difficulty, Iraqi and Afghanistani innocent people were killed and independent countries were attacked. These are not the behaviors of a human being like Mr. Bush. When Adlof Hitler came to learn that there were some human-formed aliens in Jew, without justifying aliens he decided to kill Jews at random during the World War II. But it was his guilty to kill innocent people of Jew. I think, George W. Bush is more hatred person than Adlof Hitler. Israel often attacks Palestine and kills a number of innocent people including babies, women to capture land. And we see that Human Rights Organizations remain silent. I think aliens run the human rights organizations because aliens do not understand humanity and they only know how to decrease the number of human beings. Moreover, Princess Diana and Michael Jackson were killed by the conspiracy of Aliens.

Homosexuality is one of the most identified features of the human-formed aliens. Mr. Barak Obama supported it. That means he is involved in homosexuality. The persons who practices this bad habit are sick or human-formed aliens. They want to decrease the number of human beings. It is a prerequisite condition of aliens. Aliens want to kill human beings. For this reason they encourage individuals who kill spam, do abortion and suicide, sex with others spouses and divorce etc. Aliens do not want to see husband-wife harmony. It is a matter of sorrow that aliens capture the mass media of the world by their gold power. Indirectly they are investing in the films of the Hollywood, Bollywood and so on and video songs of the famous singers for advertising their pyramid sign, one eye, two horns through fingers.

Animals

Human beings and other animals live only in the earth. Without the earth no animal can live in other planet because of CO_2O_2 cycle and the distance from the sun. There is no CO_2O_2 cycle in the universe except the earth. We, Human beings, can never naturally change physical form. We can travel any space according to our scientists' made vehicle's speed. According to the Scientist Albert Einstein we cannot travel with the speed of light. But I think if we make friendship with jinn, we can travel one planet to another planet with jinn's speed.

The earth, only one planet for living in the universe

Theory of Travelling Time in the Universe

Everything in the universe is moving. Everything has definite speed, gravity and axis way. The universe seems to expand. It seems because all things are moving with their own unique speeds. There are no two things which have same speed in the universe. The moon is used for the counting months and the sun is used for the counting years in the context of the earth. We see day night because of the earth's spinning.

To reach any planet from the earth if media of speed is more than the speed of light, it will take less time than that of light speed. Inversely it will take more time than that of light speed.

Here, Universe travelling time is equal to the product of Three quantities. They are the distance from the earth, speed media and the constant of travelling universe.

That is,

$T_t = c^{-2}DS$ seconds = DS / c^2 seconds

Where, T_t = time of travelling planet

c^{-2} = Constant of travelling universe

D = Distance from the earth

S = media of speed

c = 299729 km/second = speed of Alien = speed of light

$c^2 = (299729)^2 = 89837473441 = 8.9837473441 \times 10^{10}$

Constant of the travelling Universe

= $c^{-2} = 1/c^2 = 1/(8.9837473441 \times 10^{10})$

   = $(1/8.9837473441) \times (1/10000000000)$

= $0.11131212418353923685118788402058 \times 0.0000000001$

= $0.000000000011131212418353923685118788402058$

There are several media of speed in the universe.

Jinn or alien speed (Light speed) = c = 299792 km / second

Angel speed = as far as we see / second [approx.]

Human made various speeds
- i) Rocket speed
- ii) Plane speed
- iii) Train speed
- iv) Bus speed
- v) Car speed
- vi) Running speed
- vii) Walking speed

Various animals have various speed.

Here, Angel speed > Jinn speed > Human made media's speed > running speed > walking speed

According to Einstein's Theory of Relativity, there is an unimaginable difference between the time value of the earth and the time value of beyond the earth.

1 day of the earth = 86400 seconds and c = 299792 km/second

50000 years of the earth = 50000 X 365 X 86400 seconds
= 50000 X 365 X 86400 X 299792 km
= 472712025600000000 km
= (472712025600000000 / 9452253744000) light year

We know that,
Unit of time is second
1 minute = 60 seconds
1 hour = 60 minutes = (60 X 60) seconds
= 3600 seconds
1 day = 24 hours = (24 X 3600) seconds
= 86400 seconds

1 month = 30 days = (30 X 86400) seconds
= 2592000 s

1 year = 365 days = 86400 X 365 seconds
= 31536000 seconds
1 light year = 31536000 X 299729 km
   = 9452253744000 km
800 light year = 800 X 9452253744000 km
   = 7561802995200000 km

To sum up,
Alien speed = speed of light = c = 299729km/ second
Angel speed =7561802995200000 km / second [approx.]
Constant of the travelling Universe = c^{-2} = $1/c^2$
Tt = c^{-2}DS   seconds
Therefore, in the universe angel speed is the fastest speed of all things.

Theory of the Global Warming

At present the global warming and climate change is the most talked topic in the world. We know that global warming means the increasing average global temperature. Various scientists and researchers give the reasons, effects and probable solution of this global problem in literature manner. Recently a scientist has said that the earth is closing to the sun. For this the temperature of the earth is increasing. But no scientist or researcher develop a theory of global warming and climate change. It is I who am a student of welfare economic system since 2013 and practitioner Ecophysics (Economics + Physics) making a mathematical model for the global warming. By this model we can calculate the global warming, calculate the amount CO_2, minimize the amount CO_2 (Carbon dioxide) for the green world, determine the amount of other greenhouse gases, how many trees will need to remove global warming, determine the time when we will get the dreamy healthy world and find out the time when we control the weather of the world.

I have calculated one degree Celsius global warming is equal to (*48184380* + *a*) *kiloton CO_2 and one kiloton CO_2 is equal to 5.42 kiloton O_2 in the context of absorbing by a tree*. **One degree Celsius Global warming is equal to 66.02328T^{-1} degree adjacent angle**. In micro sense I have made the Global Warming Micro Model and then in macro sense I have established the Global Warming Macro Model. I believe that men have the power to control the weather of the world. It is our sacred duty to protect other animals and look after the earth like our sweet homes.

I think today's world is like a poor debt family in CO_2O_2 cycle. If CO_2 is Expenditure and O_2 (Oxygen) is income then the world's income (O_2) is less than CO_2. This is a problem and the problem is too large for us to imagine it.

In order to develop this model I have used velocity, acceleration and differential. In micro sense, to solve the problem we have to increase O_2 and at the same time we have to minimize CO_2. In macro sense, we have to minimize greenhouse gases (CH_4, N_2O and other green house gases) like compound interest loan and also minimize thermal energy which created by friction . This thermal energy is always ignored by scientists and researchers. I think this is a small matter but significantly big problem.

Fundamental Concepts of Global Warming Model

Global warming depends on the proportion of two groups elements. They are the increasing elements and the decreasing elements. The increasing elements are CO_2, CH_4 (Methane) , Nitrous Oxide(N_2O) and other greenhouse gases. The decreasing elements are O_2 and human's propensity to save the earth. **If the value of proportion of two groups is more than one, the global warming increases and if the value of the proportion is less than one, the global warming decreases.**

The relationship between the global warming and the increasing elements (CO_2 and other GHG) is positive but there is negative relationship between the global warming and the decreasing element (O_2). Oxygen is only hero elements to protect the earth like a hero protects a heroine from evils. As trees can absorb CO_2 which is the most responsible element for global warming and release O_2 for animals living. Besides, more than sufficient trees can help human beings control CH_4, Nitrous Oxide and other greenhouse gases. So I concentrate on CO_2O_2 cycle which is not found any other planet I think. Many scientists believe that after 2030 many animals will be extincting for global warming. So it is important to know how much O_2 require against global warming and climate change, how much CO_2 should minimize from atmosphere to save the world, how many trees should have for living in the world, how to control weather for protecting against natural disasters. It is high time we did something for protecting animals including human beings from extinction. I am trying to save the loving world by developing theory, promoting it all over the world to apply it for the peaceful world of our next generation.

Global Warming Micro Model

" **The more the value of the proportion of the efficiency of CO_2 emissions and the efficiency of O_2 production increases, the more the value of multiplier of global warming will increase and inversely decrease. When the value of the proportion is one, the Average Global Temperature will be constant in the context of CO_2O_2 cycle.** "

$w = c\,(x)^{-1}$ where, $c > 0$, $x > 0$ and if $w > 1$ then the global warming exists

$\theta = w\,(90/100)$ degree adjacent angle

Gw = Global warming = $\theta / 66.02328 T^{-1}$ °C

GW indicates the Average Global Temperature or Global Warming
c = the efficiency of CO_2 emissions
= $(CO_2 f - CO_2 i) / CO_2 f$
= $\Delta CO_2 / CO_2$ [$CO_2 f$ = final CO_2 and $CO_2 i$ = initial CO_2]
x = the efficiency of O_2 production
= $(O_2 f - O_2 i) / O_2 f$
= $\Delta O_2 / O_2$ [$O_2 f$ = final O_2 and $O_2 i$ = initial O_2]
we get c by burning fossil fuels (i.g. coal, oil and gas) and x by planting trees and the number of trees in this world. In global warming level $c > x$. The world people face this great problem now.

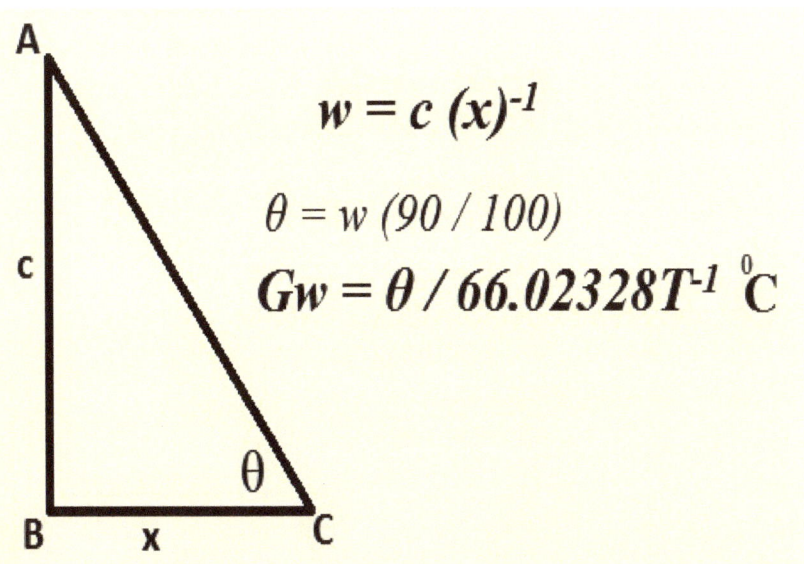

Figure: Global Warming Micro Model

In the figure, ABC is a right angled Triangle. Angle ABC = 90°, Angle ACB = θ, AB = perpendicular, BC = Base and
AC = Hypotenuse. By increasing BC the value of θ will decrease. As a result the global warming will decrease.

In 2013 the amount of $CO_2 = CO_2i$ was 35837591 kiloton and In 2014 the amount of $CO_2 = CO_2f$ was 36138285 kiloton according to The World Bank Data Group.

c = (36138285 − 35837591) kt / 36138285 kt
  = 300694 kt / 36138285 kt
  = 0.00839046 kt

Therefore, $c = \Delta CO_2 / CO_2 = 0.00839046$ kt

In 2013 the amount of forest land was 40057482 square kilometer and In 2014 the amount of forest land was 40024403.3 square kilometer according to The World Bank Data Group.

We know that a matured leafy tree produces 260 pounds O_2 in a year and this will do enough for two men according to *Environment Canada, Canada's national environmental agency. They said,"* "On average, one tree produces nearly 260 pounds of oxygen each year. Two mature trees can provide enough oxygen for a family of four."

According to NC State University, "a tree can absorb as much as 48 pounds of CO_2 per year."

If we combine two concepts in the context of a tree, we get, 48 pounds CO_2 = 260 pounds O_2
1 pound CO_2 = (260 / 48)pounds O_2
= 5.42 pounds O_2 [approximately]
Therefore, 1 kiloton CO_2 = 5.42 kiloton O_2 [approximately]
That is, If we want to remove 1 kt CO_2, we need 5.42 kt O_2. [appr.]

Suppose, T matured leafy trees are in one square kilometer
In one square kilometer,
T produces $(260 * T)/ 2204.6$ kt O_2 [since 1ton = 2204.6 pounds]
$= 0.12T$ ton O_2
So, one square kilometer forest land produces $0.12T$ kt O_2
In 2013 the world produced the amount of $O_2 = O_2i$
$= 40057482 * 0.12T$ ton O_2
$= 4806897.84T$ ton O_2
$= 4806897.84T / 1000$ kt O_2 [since 1 kiloton = 1000 ton]
$= 4806.89784T$ kt O_2
In 2014 the world produced the amount of $O_2 = O_2f$
$= 40024403.3 * 0.12T$ ton $O2$
$= 4802928.396$ T ton O_2
$= 4802928.396$ T $/ 1000$ kt O_2 [since 1 kiloton = 1000 ton]
$= 4802.928396T$ kt O_2
$\Delta O_2 = (4802.928396T - 4806.89784T)$ kt O_2
$= -3.969444T$ kt O_2
To save the earth for existing of animals we should increase the amount of oxygen but we cut down trees for our comfortable in exchange of our slowly but surely dead of our future generation.
Negative value of ΔO_2 is the evidence of that word.
Avoiding negative sign we get $\Delta O_2 = 3.969444$ kt
$x = 3.969444T / 4802.928396T$ kt O_2
$= 3.97T / 4802.93$ T kt O_2
$= 0.00082658T$ kt O_2
$= 0.00082658T$ kt $/ 5.42$ kt CO_2 [appr.] [since, 1kt CO_2 = 5.42 kt O_2]
$= 0.0001525T$ kt CO_2
We, the inhabitants of the earth, make c = 0.00839046 kt CO_2

and $x = 0.0001525T$ kt CO_2 [approximately]
We have to increase the value of x more than c. But we have increased c more than necessity. As a result, the extinction of animals including human beings count down.
$w = c/x = 0.00839046$ kt $CO_2 / 0.0001525T$ kt CO_2
$= \mathbf{55.0194T^{-1}}$ **kt** $\mathbf{CO_2}$
$w = 55.0194T^{-1} * 90$ degree $/ 100$
$= 49.51746T^{-1}$ degree adjacent angle
According to NASA in 2014 the global temperature was 0.75 degree Celsius
0.75 degree Celsius global warming $= 49.51746T^{-1}$ degree adjacent angle
Or, 1 degree Celsius global warming
$= 49.51746T^{-1}$ degree adjacent angle $/ 0.75$
$= 66.02328T^{-1}$ degree adjacent angle

So, one degree Celsius Global warming = $66.02328T^{-1}$ degree adjacent angle

When the value of adjacent angle is one, we get the balance of $CO_2 O_2$

The more the value of w increases, the more the earth will unsuitable for living place and the more animal will extinct.
In balance level, $c / x = 1$
Or, $\Delta CO_2 / \Delta O_2 = 1$
Or, $\Delta CO_2 = \Delta O_2$
If ΔCO_2 is not equals to ΔO_2, it means it does not establish balance.
If ΔCO_2 is greater than ΔO_2, it means there is global warming.
If ΔCO_2 is smaller than ΔO_2, it means the earth is suitable for animals.

The velocity of Gw is increasing because the velocity of ΔCO_2 is increasing and the velocity of ΔO_2 is decreasing.

As a result, the velocity of ΔCO_2 is more than the velocity of ΔO_2 and the velocity of VGw is increasing That is ,
$VGw = (Gwf - Gwi) / Time$,
$V\Delta CO_2 = (\Delta CO_2 f - \Delta CO_2 i) / Time$
$V\Delta O_2 = (\Delta O_2 f - \Delta O_2 i) / Time$
Therefore , $VGw = V \Delta CO_2 / V\Delta O_2$
the acceleration of Gw is a non-uniform acceleration because the value of CO_2 , O_2 and other greenhouse gases are variable.

In my global warming micro model I would not like to discuss the other Greenhouse gases such as Nitrous Oxide (N_2O), Mithene (CH_4), Chlorofluorocarbon (CFC_{12}), Hydrofluorocarbon-23 (HFC_{23}), Sulfer Hexa Fluoride (SF_6), Nitrozen Trifluoride (NF_3) because O_2 can not absorb these gases but by increasing O_2 it may reduce other greenhouse gases. So we have to minimize CO_2 on the priority basis.

According to NASA ,
In 2014 Average Global Temperature (Gw) = 0.75 degree Celsius
Average Global Temperature in 2000 according to NASA
= 0.40 degree Celsius
The velocity of global warming from 2000 to 2014
= (0.75 – 0.40) / 14
= 0.025 degree Celsius
The velocity of CO_2 from 2000 to 2014

= (36138285 kt – 24689911 kt) / 14

= 817741 kty^{-1}

The velocity of forest land from 2000 to 2014

= (40024403.3 km^2 - 40556022.3 km^2) / 14

$= -37972.786 \text{ km}^2\text{y}^{-1}$

Every year we lose approximately 37972.786 km²y⁻¹ forest land

Avoiding negative sign we get,

the velocity of forest land is 37972.786 km²y⁻¹.

Every year we lose approximately 37972.786 km²y⁻¹ forest land.

To manage deficit O_2,
We have to plant more trees.
We have to use renewable energy.
We have to avoid furniture made of wood.
Government should take Carbon tax from the factories which use fossil fuels and the individual who uses fossil fuel for driving personal car.
After managing deficit O_2 within 20 years we will reach the green world, I think. But we have not sufficient time to turn our beloved world into the green world within 20 years. We, the inhabitants of the world, are not united for minimizing carbon dioxide and increasing tree plantation. The United Nations sometimes fails to establish harmony between two conflict countries. Sometimes the United Nations remains silent to control the capitalist. However, We, the inhabitants of the earth, may vary from colour to colour, race to race, religion to religion but we all can unite for saving the world against global warming and climate change. To protect the earth against global warming we have to work unitedly by applying this theory, I think.

Calculation of how much CO_2 is responsible for *n* degree global warming

According to NASA in 2014,
the Average Global Temperature was 0.75 degree Celsius
And according to the World Bank Carbon emissions was 36138285 kiloton
So , we get,
0.75 degree Celsius Gw = 36138285 kiloton CO_2 + $GHGo$
Or, 1 degree Celsius Gw = (36138285 kt CO_2 + $GHGo$) / 0.75
Or, 1 °C Gw = (36138285 kt CO_2 / 0.75) + ($GHGo$ / 0.75)
Or , 1 °C Gw = 48184380 kt co_2 + a
   [let, a =$GHGo$ / 0.75]

As O_2 can absorb CO_2. *a can be minimized by human behaviors. So in micro sense, Global warming multiplier depends on the proportion of c and x*

One degree Celsius Global Warming is equal to
48184380 kt CO_2 + *a*
2 °C GW = 96368760 kt CO_2 + 2a

n °C GW = n (48184380 kt CO_2 + a)

The Global Warming Macro Model

In macro sense, the Creator gives human beings the power to control the atmosphere because everything of the universe is made for human beings. Global Warming depends on human activities. Two types of coefficients are created by human activities. they are increasing elements of global warming and decreasing elements of global warming.

If the proportion of the sum of the marginal propensity of increasing elements is equal to the sum of the marginal propensity of decreasing elements, the changing rate of Global Warming will be constant and establish the green world.

The global warming Macro function,
$GW = f(P) = f(CO_2, O_2, Re, GHGo, AntiGHGo, TEc)$
Where, GW = Global Warming, P = Population, CO_2 = Carbon dioxide, O_2 = Oxygen, Re = Renewable Energy, GHGo = CH_4, N_2O and other greenhouse gases, TEc = Thermal Energy caused by friction, AntiGHGo = Action against GHGo
$GW = f(P) = hP$
Therefore, $dGW / dP = h > 0$
where, P = Population of the world, h = the marginal propensity of human being's behavior
The global warming macro equation,
$GW = f(CO_2) + f(GHGo) - f(O_2) - f(Re) - f(AntiGHGo) + f(TEc)$
Part of $GW = f(CO_2) = kCO_2$
Therefore, $dGW / dCO_2 = k > 0$

Where, c = the marginal propensity of CO_2 emissions

Part of GW = f (O_2) = $-sO_2$
Therefore, dGW / dO_2 = - s   < 0
Where, x = the marginal propensity of O_2

Part of GW = f (Re) = - r RE
Therefore, dGW/ dRE = - r   < 0
Where, r = the marginal propensity of using Renewable Energy

Part of GW = f (GHGo) = g GHGo
Therefore, dGW / dGHGo = g   > 0
Where, g = the marginal propensity of other greenhouse gases

Part of GW = f (Tec) = e TEc
Therefore, dGW / dTEc = e   > 0

Part of GW = f (AntiGHGo) = - y AntiGHGo
Therefore, dGW/ dAntiGHGo = - y < 0
Where, y = the marginal propensity of antiGHGo

The global warming macro equation,
  $GW = k\ CO_2 + g\ GHGo - s\ O_2 - r\ Re - y\ AntiGHGo + e\ Tec$
0r, $\Delta GW / \Delta (CO_2+O_2+Re+GHGo+ AntiGHGo + TEc) = k + g - s - r - y + e$
   Or, $\Delta GW / \Delta P = h = \tan\theta$
If h = 45 degree then   there will establish ecological balance
But it is a matter of sorrow that today's world face h = tanθ > 45 degree . For this reason we face global warming.

When h = tanθ = 45 degree, then *(k+g+ e) / (r+s+y) = 1*
When, tan θ > 45 dgree, then *(k+g+e) / (r + s + y) > 1*

When, tan θ < 45 dgree, then $(k+g+e)/(r+s+y) < 1$

Here, $(k + g + e)$ is the sum of k,g and e
   $(r + s + y)$ is the sum of r,s and y
So we all, the people of the world, have to work for increasing the value of $(r + s + y)$ and decreasing the value of $(k + g + e)$.
Now the world faces the following diagram,

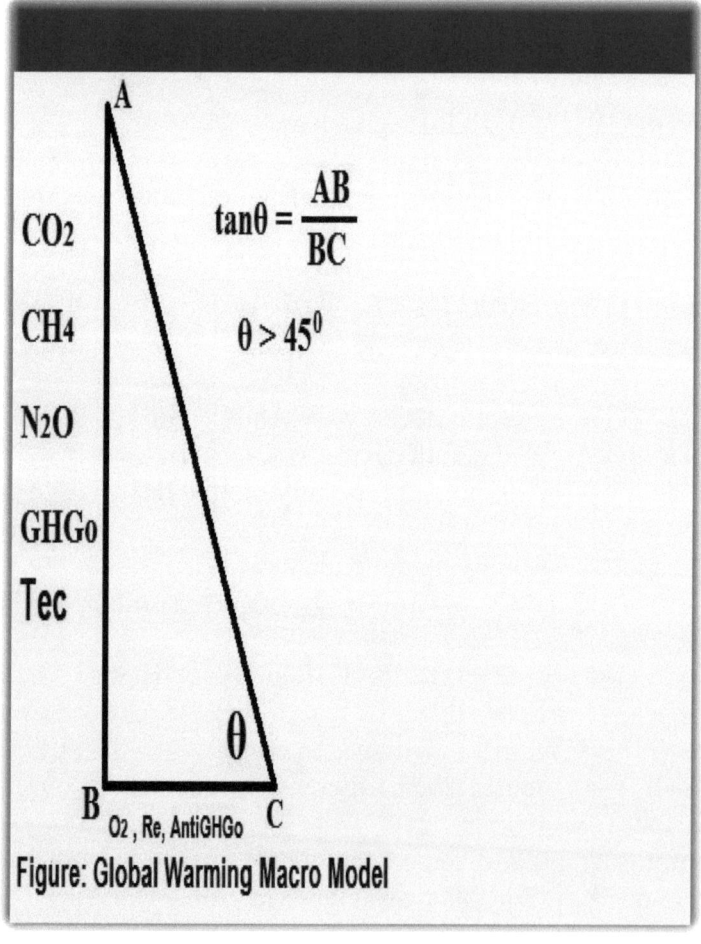

Figure: Global Warming Macro Model

In the figure, ABC is a right angled triangle. Angle ABC = 90°,
angle ACB = θ
AB / BC = tanθ = h = (k + g + e) / (r + s + y)

The 21th century of people face the gigantic problem which is called the global warming. For the global warming the world people are seeing how sea water level is increasing, polar ice is melting, many islands is sinking, many animals are disappearing and seeing various natural disasters. Suppose , you want to boil a kettle of water by heating. What will happen when the water of the kettle boils ? You can see that water level of the kettle is rising. Like the kettle our earth is heating by burning fossil fuels, using fertilizer and using various types of glass for building high rise tower, using automobiles .This heat is the main cause of increasing global temperature and increasing sea water level. When water heats up, it increases. So when the ocean warms, sea level rises. Polar ice is melting for global warming and water created by melting water runs into the ocean and thus sea water is rising gradually and it helps methane (CH_4) emit.

Many of us believe that many islands around the world are slowly but surely submerged by 2050 because of rising sea water level. This will happen only for global warming.

Some Functions for Global Warming

According to the macro model, there are some functions for global warming. I would like to discuss these functions. The first function is called carbon dioxide - global warming function.

The carbon dioxide – global warming function

Global warming depends on mainly CO_2. There are positive relationship between G_w and CO_2.
The more CO_2 increases the more global warming increases and inversely decreases.
G_w is dependent variable and CO_2 is an independent variable.
So the function is,
$Gw = f(CO_2)$
The equation of the global warming – carbon dioxide is,
$Gw = k\, CO_2$
Therefore, $\Delta Gw / \Delta CO_2 = k$

The Global Warming-Oxygen Function

There is an inverse relationship between Gw and O_2. The more the amount of oxygen increases the more the amount of Gw decreases.

The function of global warming – Oxygen is,

$Gw = f(O_2) = -s\, O_2$

Therefore, $\Delta Gw / \Delta O_2 = -s$

To minimize global warming we have to increase the quantity of forest land. Trees take CO_2 and give O_2 to maintain ecological balance. But at present the lower amount of O_2 do not perfectly absorb the higher amount of CO_2 because of human activities. We have to plant trees here and there where we find open space I think every man needs at least 226 trees to exist in the world. So we have to plant at least 226 trees per capita. By planting trees we can save many human lives and wild lives and maintain ecological balance.

The function of Carbon dioxide minimization in atmosphere by O_2

Naturally 5.42 unit oxygen is equal to 1 unit CO_2.

$$CO_2 = f(O_2) = t\, O_2$$

Or, $CO_2 / t = O_2$
Or, $CO_2\, (t)^{-1} = O_2$
Or, $(t)^{-1} = O_2 / CO_2$

Since, $CO_2 > O_2$ and in global warming level CO_2 is positive

By planting trees we will get $t = 5.42$

When we get $t = 5.42$, we will minimize $CO2$

At present t is less than 5.42. We have to face natural disasters and climate change.

The G7 (Group of Seven) countries like USA, Canada, Japan, England, France, Germany, Italy and developing counties like Chine, Brazil, India, Australia , Rasia etc use fossil fuels (i.e. coal, oil, natural gas) for increasing GNI (Gross National Income) and produce a large number of Carbon dioxide which is responsible for climate change. Most of them are unwilling to reduce the use of fossil fuels.

Due to deforestation the present number of trees can not normally absorb CO_2 from the air. If we have to eat 50 kilogram rice without any time interval, what will happen ? Just imagine. Same condition exists for the number trees of the world in the context of absorbing CO_2. Rather trees lose their normal power to absorb CO_2 and release poor oxygen.

Renewable Energy- Global Warming Function

The more the users of renewable energy increase, the more the global warming decrease. There is a negative relationship between Re and Gw. We have to use renewal energy i.g. hydro power, biomass, solar panel, wind power, biofuel etc.

$G_W = f(Re) = -rRe$
$\Delta G_W / \Delta Re = -r$ where, $r > 0$, it indicates the marginal propensity to use renewable energy.

Greenhouse Gases – Global Warming Function

The more GHGo increases, the more the global warming increases. There is a positive relationship between G_W and GHGo
$G_W = f(GHGo) = gGHGo$
$\Delta G_W / \Delta GHGo = g$ where, g is the marginal propensity of GHGo

Global Warming -Thermal Energy caused by friction Function

The more the thermal energy caused by friction increases, the more the global warming increases. There is a positive relationship between GW and Tec
$G_W = f(Tec) = e\,TEc$
$\Delta GW / \Delta TEc = e$ where, e = the marginal propensity of thermal energy

AntiGreenhouse Gases – Global Warming Function

The more AntiGHGo increases, the more the global warming decreases. There is a negative relationship between G_W and AntiGHGo

$G_W = f(AntiGHGo) = y\,AntiGHGo$
$\Delta G_W / \Delta AntiGHGo = y$ where, y is the marginal propensity of AntiGHGo

Who Are Responsible for Global Warming

Human beings create global warming. Nature have no power to create global warming. We, the human beings, are responsible for increasing global warming. We are decreasing the number of trees per square kilometer for making furniture, building living places and manufacturing various products from trees. As a result , we get low amount of oxygen from air but we produce carbon dioxide more than necessary in air. The lower capital than necessary begets the lower production. The lower production than necessity are responsible for inflation. Like this way we begets global warming by getting lower amount of O_2 and releasing more CO_2 in air. How many trees should need for a square kilometer land according to the density of population is unknown to us. I think it is time to calculate the number of trees in a square kilometer by research. It is not necessary to travel another planet like The Mars. It is necessary to find out how many trees we need a square kilometer land to get enough oxygen. Lack of oxygen we produce Carbon dioxide more than necessary.

How to Reduce the Global Warming

To reduce global warming we have to increase the marginal propensity to plant trees and use renewable energy so that CO_2 can not increase. If we use more fossil fuels we will increase more CO_2. So we have to minimize CO_2. There is positive relationship between CO_2 and global warming. If CO_2 increases, global warming will increase. We know that a matured leafy tree produces 260 pounds O_2 in a year and this will do enough for two men according to *Environment Canada, Canada's national environmental agency. They said,"* "On average, one tree produces nearly 260 pounds of oxygen each year. Two mature trees can provide enough oxygen for a family of four."

According to NC State University, *"a tree can absorb as much as 48 pounds of CO_2 per year."*

If we combine two concepts, we get, 48 pounds CO_2 = 260 pounds O_2
1 pound CO_2 = (260 / 48) pounds O_2 = 5.42 pounds O_2
Therefore, **1 kiloton CO_2 = 5.42 kiloton O_2**
That is, If we want to remove 1 kt CO_2, we need 5.42 kt O_2.
In 2018 the number of world population is 7.594×10^9 according to The World bank Data Group.
O_2 needs in 2018 for the whole population

= 7.594 x 10^9 x 130 pounds
= 987.22 x 10^9 pounds
= (987.22 x 10^9) / 2204.6 ton
= 447800054.43 ton O_2
= (447800054.43 / 1000) kt O_2
= 447800.054 kt O_2

Suppose, for all animals O_2 need in 2019 ,
 4 times x human beings required O_2
= 4 x 447800.054 kt O_2
= 1791200.22 kt O_2
= (179200.22 / 5.42) kt CO_2
= 330479.74 kt CO_2

But in 2014 we, the inhabitants of the earth, produced 36138285 kt CO_2 according to the World Bank Data Group. This amount is increasing day by day. In 2019 the earth need 330479.74 kt CO_2 but we produced 36138285 kt CO_2 in 2014 by cutting down trees, using fossil fuels (i.e. coals, gases, oils) . Extra amount we produced 5 years ago (36138285 – 330479.74) = 35807805.26 kt CO_2 which was **108.35 times more.** The gigantic problem is this amount CO_2.

On the other hand, the world has only 39958245.9 square kilometer forest land. How many trees are there in a square kilometer forest land ? We don't know. But we know we are minimizing forest land by cutting trees for making furniture, using industrial purposes etc.

Suppose , T matured leafy trees are in one square kilometer. The number of trees in the world in 2016 is 39958245.9 x T trees.

They produced 3995845.9 T (260 / 2204.6)   ton
= 3995845.9 T x 0.12 ton
= 479501.508T ton O_2
= 479501.508T / 1000 kt O_2

$= 479.502T$ kt O_2

The Green World

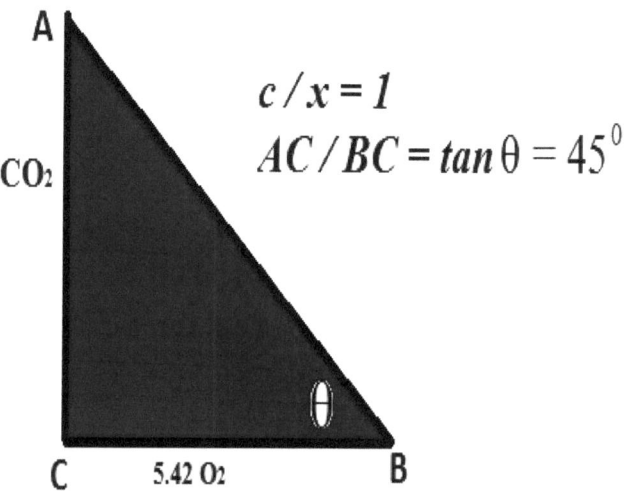

Figure : The Green World

In green world level, $CO_2 / 5.42\ O_2 = 1$
In the figure, According to the right angle triangle ABC,
$\tan \theta = AC / BC = CO2 / O2 = 1 = 45$ degree
In the green world level, everywhere we will find trees and trees. The world will seem to be a big jungle. The inhabitants of the world will overcome natural disasters and live happy naturally. The proportion of the marginal propensity of using fossil fuels and the tree plantation will same, there have no extinction of any animal. There will establish ecological balance.

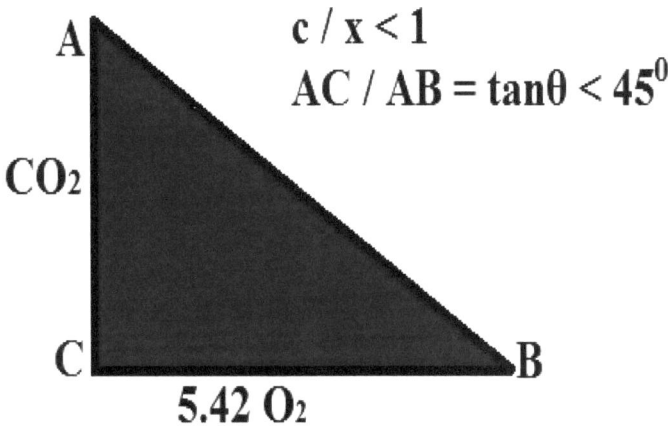

Figure: The Healthy Dreamy World

When , the world reach $CO_2 / 5.42\ O_2 < 1$,
then the world will be called Dreamy Healthy World. I think every man needs at least 226 trees to exist in the world. So we have to plant at least 226 trees per capita. By planting trees we can save many men and women and wild lives and maintain ecological balance.

.

Turning the Present World into the Green World

The present world will be unsuitable for living by 2050. Many animals will extinct by this time. So it is high time we, the people of the world, increased the marginal propensity to produce oxygen and decreased the marginal propensity of carbon emissions so that we can turn the present world into the green world. According to my model when we reach c = x , we will make the existing world the green earth.

For increasing the number trees we have to take the following steps.

The marginal propensity to plant trees should be greater than the marginal propensity to cut down trees.

that is, $p > d$ Where, $p, d > 0$

p = the marginal propensity to plant trees ,

d = the marginal propensity to cut down trees

if $p > d$ then the world people can increase the number of trees . that means $x = p - d$ where, $x > 0$

If p < d , then the world people can decrease the number of trees for their extinction.

That means $x = p - d$ where $x < 0$ which we the world people have created. For this we are facing global warming. To save the world against global warming we should increase the value of x .

The United Nations, World Bank and climate related organizations should encourage the world people to plant trees. The botany experts should research for knowing how much O_2 (oxygen) produce per tree. We should plant the tree which produces the largest amount of oxygen .

To save the world against the global warming we should decrease the value of w. To minimize the amount of carbon dioxide we have to use more renewable energy than fossil fuels.

If the marginal propensity to use fossil fuels is less than the marginal propensity to use renewable energy, we will decrease the value of w.

Besides,

We should be aware of using the things which produce the greenhouse gases.

The Botany experts should find out how many trees need for one square kilometer forest land and find out the tree that produce more oxygen than normal tree.

We always remember that space travelling is less important than tree plantation. The earth is the only place for animals living because $CO_2 O_2$ cycle exists only this planet and the central point of the universe is here. I have discovered the central point of the universe by my Universal Model. Finally I would like to say " First save the earth against global warming then research another planets.

Analysis of the collected data

Year	Population (Billion)	ΔP	Forest Area (Sq.km)	ΔF	CO₂ Emission (Kiloton)	ΔCO₂	GW (°C)
1999	6.035	-	40628689.7	-	24059187	-	
2000	6.115	0.080	40556022.3	-72667.4	24689911	630724	0.40°
2001	6.194	0.079	40510303.3	-45719	25276631	586720	
2002	6.274	0.08	40464584.1	-45719.2	25646998	370367	
2003	6.353	0.079	40418865.2	-45718.9	27047792	1400794	
2004	6.432	0.079	40373146	-45719.2	28393581	1345789	
2005	6.513	0.081	40327427	-45719	29490014	1096433	
2006	6.594	0.081	40293287.6	-34139.4	30568112	1078098	
2007	6.675	0.081	40259147.7	-34139.9	31180501	612389	
2008	6.758	0.083	40225008.7	-34139	32181592	1001091	
2009	6.841	0.083	40190869.1	-34139.6	31891899	-2829693	
2010	6.923	0.082	40156729.7	-34139.4	33472376	1580477	
2011	7.004	0.081	40123639.2	-33090.5	34847501	1375125	
2012	7.087	0.083	40090560.5	-33078.7	35470891	623390	
2013	7.171	0.084	40057482	-33078.5	35837591	366700	
2014	7.256	0.085	40024403.3	-33078.7	36138285	300694	0.75°
		ΣΔP = 1.221		ΣΔF = -604286.4		ΣΔCO2 = 9539098	

Data source :
World Bank Data Group (Population, Forest Area and CO2 Emissions)   and NASA(Global warming)

ΔP = current year – previous year,
ΔF = Current year – previous year ,
ΔCO₂ = Current year – previous year
The average value of ΔP from 2000 to 2014
= ΣΔP / n = 1.221 / 15 = 0.814

The average value of ΔF from 2000 to 2014 = ΣΔF / n = - 604286.4 / 15 = - 40285.76 sq.km. The average value of ΔCO_2 from 2000 to 2014 = $\Sigma \Delta CO_2$ / n
= 9539098/15 = 635939.87 kt

Comment: there is positive relationship between Population and Carbon dioxide but negative relationship between population and forest area. So lack of sufficient oxygen the world temperature is increasing gradually.

The velocity of CO_2 emissions from 2000 to 2014 (VCO_2)
= (36138285 – 24689911) / 15
= 763224.93 kt
The velocity of forest area from 2000 to 2014(VF)
= (40024403.3 – 40556022.3) / 15
= - 531619 square land
Suppose, the number of trees per square km = *T*
According to Canada Environment Agency ,
1 matured tree produces O_2 = 260 pounds.
According to them,
A tree produces O2 = 260 / 2204.6 = 0.12 ton
We know that 1000 ton = 1 kiloton
T trees produce O_2 per square kilometer = 0.12*T* ton
The velocity of O_2 from 2000 to 2014 (VO_2)
= VF x 0.12*T* ton = - 531619 x 0.12 *T* ton
= - 63794.28 *T* ton
= - 63794.28 *T* / 1000 = - 63.79*T* kt
Every year we lose O_2 63.79 *T* kt (approximately)
So we, the inhabitants of the earth, should produce 2 times of 63.79 *T* kt O_2 per year for saving the world.
The velocity of the global warming from 2000 to 2014 (VGw) = (0.75 – 0.40) / 15 °C = 0.023 °C

The velocity of Population growth = $(7.256 \times 10^9 - 6.115 \times 10^9) / 15 = 1.141 \times 10^9 / 1 = 76066666.67$

From the above discussion, It can be said that in nature there are negative relationship between global warming and forest areas which produce O_2 for maintaining ecological balance. **So only through increasing forest area global warming can be minimized.**

O_2 inhaled from air in 2014

= (Total forest area x 0.12 T) / (7.256 x 10^9) ton per person

= (40024403.3 X 0.12 T) / (7.256 x 10^9) ton per person

= 4802928.396 T / 7256000000 ton per person

= 0.00066 T ton / person

CO2 exhaled in air in 2014 year

= 36138285 / (7.256 x 10^9) kt/ person

= (3.6138285 x 10^7) / (7.256 x 10^9)

= 3.6138285 / 7.256 X 10^2

= 3.6138285/ 725.6

= 0.0049805 kt/ person

= 4.9805 ton/ person

According to NC State University, a tree can absorb as much as 48 pounds of CO_2 per year.

So, 48 pounds = 48 /2204.6 ton = 0.022 ton (approximately)

The number of trees is needed for absorbing CO_2

= 4.9805 / 0.022 trees per person

= 226 trees (Approximately)

A human being should plant 226 trees (approximately) for saving the world against climate change and global warming.

According to World bank Data Group, in 2000

Carbon emissions 24689911 kt,
Methane (CH_4) Emissions 6480650 kt (equivalent CO_2),
Nitrous oxide emissions 2920510 kt (equivalent CO_2)
According to NASA, Global Warming $0.40°$ in 2000
So, $Gw = CO_2 + CH_4 + N_2O + GHGo$
$0.40°\, Gw = (24689911 + 6480650 + 2920510)\, kt + GHGo$
Or, $0.40°\, Gw = 34091071\, kt\, CO2 + GHGo$
Or, $1°\, Gw = 85227677.5\, kt\, CO2 + (GHGo / 0.40°)$
Or, $1°\, Gw = 85227677.5\, kt\, CO2$ [$(GHGo / 0.40°) \approx 0$]
So, $1°\, Gw = 85227677.5\, kt\, CO2$ (approximately)

It is high time we planted at least 226 tree plants wherever we like for saving our next generation like my only son Shayan against global warming and climate change.

Result of the Theory

In micro model,

$w = c\,(x)^{-1}$ where, $c > 0$, $x > 0$ and if $w > 1$ then the global warming exists

$\theta = w\,(90/100)$ degree adjacent angle

Gw = Global warming = $\theta / 66.02328 T^{-1}$ °C

1 pound CO_2 = 5.42 pounds O_2 [approximately]
1 kiloton CO_2 = 5.42 kiloton O_2 [approximately]
If we want to remove 1 kt CO_2,
we need 5.42 kt O_2. [approximately]

Suppose, T matured leafy trees are in one square kilometer one square kilometer forest land produces $0.12T$ kt O_2

one degree Celsius Global warming = $66.02328T^{-1}$ degree adjacent angle

one °C global warming is equal to ($48184380 + a$) kiloton CO_2

Macro model,

$GW = hP$

$= KCO_2 + g\ GHGo + eTEc - sO_2 - r\ RE - y\ AntiGHGo$

Global Warming depends on population behavior.

To remove global warming every person needs to plant 226 trees for saving the world and our future generation. Otherwise, you may become the richest person in the world but you will not be able to save your life against global warming.

I think, by applying this model we, the inhabitants of the world, can minimize 0.30 °C global warming within 5 years and within 20 years we will get the healthy dreamy world.

www.ingramcontent.com/pod-product-compliance
Lightning Source LLC
Chambersburg PA
CBHW040325220526
45473CB00009B/2575